EARTH'S DETECTIVES

ALL ABOUT MINERALS

BY REBECCA STORM

CONTENTS

WHAT ARE MINERALS?	4
HOW DO MINERALS FORM?	6
MINERAL SHAPES	8
MINERAL COLOURS	10
HOW HARD ARE MINERALS?	12
HOW WE USE MINERALS	14
MINERAL DETECTIVES	16
AMAZING MINERAL PLACES	18
HOW TO FIND MINERALS	20
ROCKY MINERALS	22
MINING FOR MINERALS	24
DISPLAY-WORTHY MINERALS	26
RECORD-BREAKING MINERALS	28
TRUE OR FALSE?	30
GLOSSARY & INDEX	32

Words in **BOLD** can be found in the glossary.

Copyright © 2025 Hungry Tomato Ltd

First published in 2025 by Hungry Tomato Ltd
F15, Old Bakery Studios, Blewetts Wharf, Malpas Road, Truro, Cornwall, TR1 1QH, UK.

No part of this publication may be reproduced, stored in a retrieval system, or transmitted in any form or by any means, electronic, mechanical, photocopying, recording, or otherwise, without prior written permission of the copyright owner.

A CIP catalogue record for this book is available from the British Library.

ISBN 9781835690819

Printed in China

Discover more at
www.hungrytomato.com

Picture credits:
Abbreviations: m-middle, t-top, l-left, r-right, bg-background.

iStock: 4ml. Shutterstock: 4b, 4mr; 3DML 20bl (rock pick); Abdul Matloob 27tl; Africa Studio 10tr; Albert Russ 26mr, 28ml; Aleksandr Pobedimskiy 13(corrundum); Aleksey Matrenin 21mr; Amanda Mohler 9tr; Arcady 21br; Bashi Kikia 18b; Bjoern Wylezich 9br, 22tl, 24tl; BrankoG 25tl; Breck P. Kent 28br; Cagla Acikgoz 26tl; COULANGES 25bl; Dani-G 31tr; David Tadevosian 11br; Dean Drobot 21tl; DiamondGalaxy 29mr; Dianne Wickenden 19tl; DmitrySt 8tr, 13(diamond); Dmivmas 23tl; ENRIQUE ALAEZ PEREZ 1tr; fanjianhua 10bl;Fokin Oleg 12(fluorite); iadams 20b (magnifying glass); Idutko 13tl; Ihor Bondarenko 11ml; IM Imagery 15mr(battery); ivan_kislitsin 15bl; Jaroslav Moravcik 31ml; Julia Reschke 14tl(gold), 24bl; Kei Shooting 5br; KrimKate 12(calcite), 12(apatite), 23mr; Kwin_New 14tl; luchschenF 16m; Marlene Gracia 21ml; Michael LaMonica 13tr, 28mr; milicenta 17b; Minakryn Ruslan 9ml, 23bl, 24mr, 27mr, 27bl, 29tl; mykolastock 20b (camera and rucksack); New Africa 30tr; Nowamhere 14b(steel); Nur Ismail Photography18t; Olga Gavrilova 7t; Owlie Productions15mr; Ozgur Coskun FC; Parnarat thammachanon 5t; Peter Hermes Furian 8br; Perekotypole 1bg, 19b; photo-world 14tl(blue sapphire); Prostock-studio 31br; Sebastian Janicki 6tr, 8bl, 13(quartz), 13(feldspar); TR_Studio 13(topaz); Tyler Boyes 8mr, 13tl(graphite); RHJPhtotos 15m (cobalt), 25mr; Roschetzky Photography 114br; Wirestock Creators 12(gympsum), 19tr; worradirek 17t; Vadim Sadovski 30b; vlue 2-3bg; vvoe 7bl, 12(talc), 14br(hermatite), 22bl, 22mr, 26bl; Yvonne Baur 6br; Zelenskaya 15b(copper).

Every effort has been made to trace the copyright holders, and we apologise in advance for any unintentional omissions. We would be pleased to insert the appropriate acknowledgements in any subsequent edition of this publication.

WHAT ARE MINERALS?

Minerals are chemicals that occur naturally in the ground. There are many different types of minerals, and they come in lots of different shapes and colours.

Minerals are the basic parts from which rocks are made. All rocks are made of at least one mineral, but most contain two or more.

Some minerals can be **SOFT AND POWDERY...**

some minerals can be **SHARP AND SHINY...**

some minerals **JOIN TOGETHER TO FORM ROCKS.**

Minerals are very important for **PLANTS, ANIMALS, AND EVEN US!** We need the mineral calcium, which we get from milk, for our bones to be strong. The salt you have in your food is a mineral, too.

We use minerals to make things as well. Some of the most common minerals are metals, like gold and silver. There are minerals in toothpaste, paint, and even in computers! You probably use things with minerals in all the time!

Without minerals, we wouldn't be able to build the **circuit boards** that **ELECTRONICS NEED TO WORK!**

SALT comes from huge lumps of minerals that have been ground down into tiny **grains**.

HOW DO MINERALS FORM?

Minerals can form in many different places. Although most form deep underground, some form on the Earth's surface.

Minerals are made of chemicals. These chemicals are often found in liquids. When the liquids are heated or cooled down, the chemicals in them harden and join to form minerals.

Minerals occur naturally. They are inorganic, which means that they don't come from animals or plants.

> Citrine is made of the chemicals silicon and oxygen, and often forms as a crystal.

Most minerals form deep underground where it's so hot that rock melts, and the chemicals inside the rock **dissolve**. When the **molten** rock cools, the chemicals harden to form new minerals and rocks.

> Sometimes molten rock bursts out of a volcano and cools down on Earth's surface.

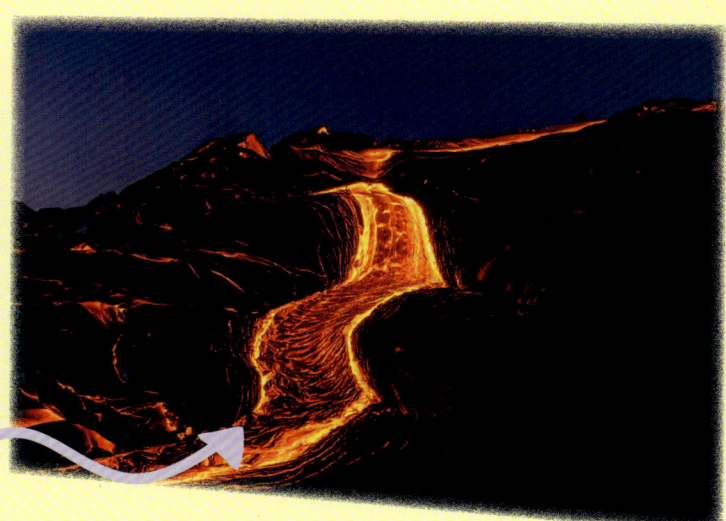

Some minerals form on the surface of the Earth, mostly when water **evaporates** and leaves minerals behind. This is the Uyuni Salt Flat in Bolivia. When water evaporates here, salt is left behind.

FROM MINERAL TO ROCK:

Rocks form when minerals join together. When rocks form from hot, molten rock, they are called **igneous rocks**.

Sometimes extreme pressure under Earth's surface changes mineral shapes, which forms new types of rocks. Rocks made this way are called **metamorphic rocks**.

Diorite is an igneous rock that's made of the minerals feldspar and hornblende or biotite.

MINERAL SHAPES

We can identify minerals by looking at their shape. The chemicals within minerals form in repeating patterns, which gives them a crystal-like shape and unique appearance. Crystals have smooth faces and straight edges.

SHAPE IS EVERYTHING

Diamond and graphite are both made of carbon, but because the carbon joins in a different pattern, the two look very different! Minerals that are made of the same things but have a different crystal structure are called polymorphs.

DIAMOND

GRAPHITE

SYMMETRICAL SHAPES

Some crystal shapes are symmetrical. This means they look the same from all sides.

MINERAL QUARTZ

It has crystals that form in long, six-sided columns.

TABLE SALT

You have to use a microscope to see them in detail, but the crystals of table salt form in cubes.

NON-SYMMETRICAL SHAPES

Not all crystal shapes are symmetrical. These are often part of mineral displays because of their unusual shapes.

GYPSUM

This is a type of gypsum. It's called "desert rose" because it's shaped like rose petals and forms in the desert.

COPPER

The mineral copper forms in shapes that look like a plant! This type of shape is called a "dendritic" shape. "Dendritic" means "tree-like".

HAEMATITE

Haematite often forms in round, bubbly shapes! The crystals that make these shapes are so small you can't see them with the naked eye.

MINERAL COLOURS

Minerals can be many different colours. A mineral's colour can help you identify which type of mineral it is. Colourful minerals have been used in different ways for thousands of years.

MINERAL MAKE-UP

Many cultures have used powered minerals in make-up and skin treatments. Even up to the 1900s, people used products containing high levels of arsenic and mercury. They didn't know that these minerals are **poisonous**!

MINERAL DYE

For thousands of years, minerals, such as lazurite, have been used to make **dye**. Ancient cultures all over the world used powered minerals to make dyes that would colour their clothes.

MINERAL PAINT

In the past, minerals were used to make paint. The paint was made by crushing the minerals and mixing them with animal fat or oil to make a paste.

The first paints, found in cave art from thousands of years ago, used basic colours that would have been the easiest to make, including black, white, and brown.

The ancient Egyptians used lots of bright colours. They made blue from the minerals lapis lazuli and azurite, green from malachite, and yellow from orpiment.

Today, we know that many of the minerals used in ancient paints are poisonous. Scientists have found other ways to create **pigments** that are safer.

HOW HARD ARE MINERALS?

While some minerals are soft, others are really hard. Scientists invented a scale to measure how hard different minerals are. It runs from 1, which is the softest, to 10, which is the hardest. It's called the **Mohs Hardness Scale**.

HOW DOES IT WORK?

The scale shows hardness through how easily a mineral can be scratched. If you can scratch something with your fingernail, it is softer than 2 ½ on the hardness scale.

TEST IT OUT!

You can also use another mineral to do a hardness test. A hard mineral will scratch a softer one.

Talc is the softest mineral. It breaks very easily into thin silver or white flakes. It's most familiar form is talcum powder.

Diamond is not just the hardest mineral, it's one of the hardest naturally-occurring materials! It's also one of the most expensive.

6. **7.** **8.** **9.** **10.**

FELDSPAR 6

QUARTZ 7

TOPAZ 8

CORUNDUM 9

DIAMOND 10

HOW WE USE MINERALS

We can use minerals to make lots of amazing and useful things. Did you know all these everyday things are made from minerals?

SPARKLING JEWELLERY

Gold and silver are minerals that are commonly used to make jewellery. Gemstones, such as rubies, sapphires, and amethysts, are minerals that have been cut and polished. They are perfect in jewellery.

STRONG CONSTRUCTION

The mineral haematite is made into iron. Under high temperatures, iron can be turned into steel which is really strong. Steel is used to build bridges, buildings, cars, and many more things.

USEFUL PENCILS

We wouldn't have pencils without graphite - that's the dark material inside. Graphite is one the softest minerals on Earth, which is why it makes marks when you write or draw.

POWERFUL BATTERIES

Lots of machines need batteries to work, including cars and computers. It takes lots of minerals to make batteries, such as cobalt, lithium, and graphite.

ELECTRICAL POWER

Chalcopyrite is a type of copper mineral that is used to make electrical wires, as well as power tools and pipes. People have used copper for thousands of years.

MINERAL DETECTIVES

There's so much to learn about minerals, but how do we know about them and the secrets they hold?

The scientists who study minerals are called mineralogists. It's thanks to their studies that we have learnt so much about minerals, their unique features, and all the things they can do.

Mineralogists spend lots of time in the **LAB**, doing experiments on rocks and minerals to identify them, and see what they're made of.

They use high-tech machines to learn what chemicals are inside the minerals, how they react to other materials, and what their structure is like. This helps them learn about new types of minerals.

Some mineralogists work at extraction sites. This fieldwork can include collecting samples, advising people how to extract minerals from different materials without damaging them, and reporting data to help create a mineral map of a particular area.

Mineralogists do an important job: their discoveries increase our understanding of Earth and the processes that have been happening for millions of years!

WHO KNOWS WHAT THEY'LL DISCOVER NEXT?

AMAZING MINERAL PLACES

Minerals are found all over the world, and make up most of our natural landscapes. Here are some incredible places that are rich with minerals!

Eldhraun lava field, Iceland

Lava is full of minerals like magnesium and iron. Located in Iceland, the Eldhraun lava field is one of the biggest lava flows in the world. It is 250 years old, and is now covered in moss.

Jwaneng diamonds, Africa

The diamond mine of Jwaneng in Botswana, Africa, is one of the richest in the world. It usually produces between 12 and 15 million **carats** of diamonds every year!

Goldfields, Australia

Australia is known for being an amazing place to find gold. While there are many large **mines**, people have found **gold nuggets** with metal detectors in the **outback**'s goldfields!

Pulpí Geode, Spain

Pulpí Geode, which is filled with huge gypsum crystals, is one of the largest crystal caves ever found, and is the biggest geode in the world. It was discovered in 1999 in an old silver mine.

Rainbow Mountain, Peru

Also known as Vinicunca, Peru's Rainbow Mountain is made of 14 different minerals, all bright and visible in thick layers. For centuries it was hidden under snow - it wasn't discovered until 2013!

HOW TO FIND MINERALS

You don't have to be an expert to find rocks and minerals; you just need a few basic tools and to know what to look for!

TO GET STARTED, YOU WILL NEED:

- A strong backpack for carrying your tools and finds
- Old towels and plastic bags to protect your finds
- Goggles to protect your eyes
- A rock hammer
- A chisel for splitting rocks
- A magnifying glass
- A camera for taking pictures of minerals that are too big to carry
- A notebook and pencil for writing down your findings

BEFORE YOU GO

Plan your trip carefully. Some places have rules about collecting minerals and rocks – make sure you know the rules where you're going so you can follow them.

STAYING SAFE

Go with an adult and stick together. Pack food, water, and a mobile phone in case you need to call for help.

Don't hammer cliffs or rock faces; only study minerals already on the ground.

FINDING MINERALS

Mineral veins running through rocky hills may be full of minerals like quartz. Because rocks are made of minerals, you can also search the beach. Always ask an adult to split a rock if you want to see inside.

DISPLAYING YOUR MINERALS

Clean your minerals carefully with a soft brush. Choose a display box with a clear front and use a piece of card to write the name of each mineral, and the place, and date they were found.

Rocky Minerals

As we have seen, there are lots of different minerals in the world. Here are some of the most common ones that make up lots of rocks.

Feldspar

- Colour: White, pink, or grey
- Hardness: 6
- Found in: Igneous and metamorphic rock

Quartz

- Colour: Mainly clear, but can be pink, white, purple, or brown
- Hardness: 7
- Found in: Igneous, metamorphic, and **sedimentary rock**

Mica

- Colour: Black, white, green, or red
- Hardness: 2½
- Found in: Igneous, metamorphic, and sedimentary rock

Olivine

- Colour: Green
- Hardness: 7
- Found in: Igneous rock

Calcite

- Colour: Clear or white
- Hardness: 3
- Found in: Igneous, metamorphic, and sedimentary rock

Garnet

- Colour: Mainly clear, but can be orange, pink, green, black, or brown
- Hardness: 7
- Found in: Igneous and metamorphic rock

MINING FOR MINERALS

Mining is the process of digging useful materials from the ground. Both rocks and minerals are mined in large quantities and used in many different things. Here are some of the most commonly mined minerals.

SILVER

- Colour: Silver
- Hardness: 2½
- Found in: Igneous and metamorphic rock

COPPER

- Colour: Orange and reddish-brown
- Hardness: 3
- Found in: Igneous and sedimentary rock

GOLD

- Colour: Gold
- Hardness: 2½
- Found in: Sedimentary rock

GALENA

- Colour: Grey and silver
- Hardness: 2½
- Found in: Igneous, metamorphic, and sedimentary rock

PETALITE

- Colour: Mainly clear, but can be pink, grey, yellow, or white
- Hardness: 6
- Found in: Igneous rock

SIDERITE

- Colour: Brown
- Hardness: 4
- Found in: Sedimentary rock

25

DISPLAY-WORTHY MINERALS

Minerals can come in all shapes, sizes, and colours. Here are some of the most stunning examples that would be perfect in a display case.

CHRYSOCOLLA
- Colour: Blue or green
- Hardness: 2-4
- Found in: Igneous and sedimentary rock

RHODOCHROSITE
- Colour: Red, pink, orange, or brown
- Hardness: 4
- Found in: Igneous and metamorphic rock

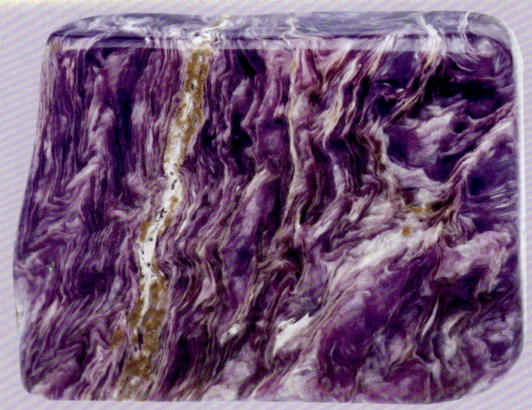

CHAROITE
- Colour: Purple with grey, white, black, or brown specks
- Hardness: 5-6
- Found in: Metamorphic rock

OPAL

- Colour: Any and all colours
- Hardness: 6
- Found in: Igneous and sedimentary rock

AZURITE

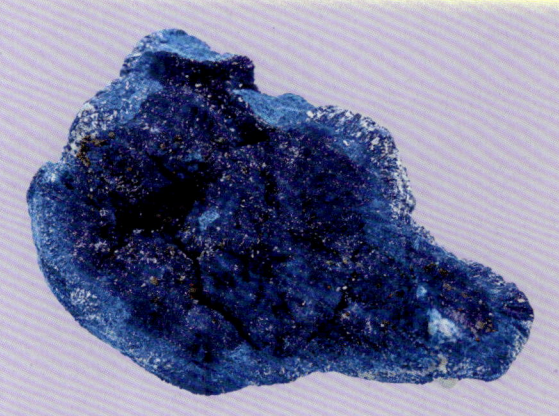

- Colour: Blue
- Hardness: 4
- Found in: Igneous, sedimentary rock, and mineral veins

COBALTOCALCITE

- Colour: Pink, red, and purple
- Hardness: 3
- Found in: Sedimentary rock

RECORD-BREAKING MINERALS

All minerals are awesome, but these minerals are the most impressive of all.

Many of the world's minerals are **DANGEROUS**! Cinnabar contains high levels of mercury, which can be poisonous if touched or breathed in. Asbestos is a group of minerals that are **fibrous** and very harmful to breathe in.

Cinnabar

Cinnabar was used for thousands of years in paint and tattoo ink!

Asbestos

Asbestos is often used in construction, but has now been banned in many countries.

Mesolite looks really **FLUFFY**, with its hundreds of hair-like needles spreading out into domes. These needles break very easily, which makes finding perfect mesolite crystals very rare.

The **OLDEST MINERAL** from **Earth's crust** is zircon, found in Australian sedimentary rock. It's thought to be around 4.4 billion years old – that's almost as old as Earth itself! The oldest zircons found were only tiny crystal grains within a large rock.

This is a zircon mineral stone.

The Cullinan diamond is the **BIGGEST GEM DIAMOND** ever found. It was discovered in South Africa, and was about the size of a grapefruit!

Naturally-formed ice is a mineral, even though water isn't! It's the **MOST COMMON** mineral on Earth's surface.

29

TRUE OR FALSE?

There is lots to learn about minerals. How well do you know your mineral facts?

On the Mohs Hardness Scale, talc is 1 and diamond is 10!

TRUE! This is a simple way for us to compare minerals, but it can be confusing – diamonds aren't 10 times harder than talc, they are 1,500 times harder!

Minerals can only be found on Earth!

FALSE! Scientists have discovered around 300 minerals in **meteorites**! About 40 of these minerals have only ever been found in meteorites – never naturally-occurring on Earth.

Not all minerals are as they seem!

TRUE! The mineral pyrite is often called "Fool's Gold" because it looks so similar that it can easily be mistaken for gold!

Minerals have only just started being used to make jewellery!

FALSE! The ancient Egyptians were one of the earliest cultures to make jewellery and other precious things from gold.

Minerals aren't just found in rocks!

TRUE! Your bones and teeth are made from minerals too, including calcium, phosphate, and hydroxide.

31

GLOSSARY

Carats – a way of measuring the weight or size of a gemstone, such as a diamond.

Circuit boards – boards which are covered in electric connections and used in many machines.

Dissolve – when a solid turns into a liquid.

Dye – a substance that can change the colour of things like clothes and hair.

Earth's crust – the very top layer of Earth's rocky surface.

Evaporates – when a liquid turns into a gas.

Fibrous – something that is made of lots of little fibres.

Gold nuggets – solid lumps of gold.

Grains – small, hard particles.

Igneous rocks – rock that form when molten rock cools down and hardens.

Metamorphic rocks – rocks that forms when heat or pressure causes rocks to change their structure or mineral composition.

Meteorites – rocks that come from space and fall to Earth's surface.

Mineral veins – distinct layers or stretches of minerals that run through rocks.

Mines – areas within the Earth that have been dug out by humans, for the purpose of finding and collecting minerals and materials deep underground.

Mohs Hardness Scale – a scale which is used to measure the hardness of minerals.

Molten – another word for melted.

Outback – an isolated and remote in-land place, often associated with Australia.

Pigments – a powdered substance that is mixed with liquid to create something that can colour other things. Pigments are used in paints and inks, and come in different colours.

Poisonous – something that can cause harm if touched, eaten, or breathed in.

Sedimentary rock – rocks that form when small, worn off pieces of other rocks become joined together in layers.

INDEX

A
Ancient Egyptians 11, 31
Apatite 5
Asbestos 28

B
Biotite 7

C
Calcite 12, 23
Calcium 5, 31
Cinnabar 28
Citrine 6
Copper 9, 12, 15, 24
Corundum 13
Crystals 6, 8-9, 19, 28-29

D
Diamond 8, 13, 18, 29, 30

E
Eldhraun, Iceland 18

F
Feldspar 7, 13, 22
Fluorite 12

G
Gemstones 14, 29
Gold 5, 14, 19, 24, 31
Goldfields, Australia 19
Graphite 8, 15
Gypsum 9, 12, 19

H
Haematite 9, 14
Hornblende 7
Hydroxide 31

J
Jwaneng, Botswana, Africa 18

M
Mesolite 28
Meteorites 30
Mineral veins 21, 27
Mineralogists 16-17

P
Phosphate 31
Poisonous minerals 10-11, 28
Pulpí Geode, Spain 19
Pyrite (Fool's Gold) 31

Q
Quartz 8, 13, 21, 22

R
Rainbow Mountain, Peru 19
Rocks
  Igneous 7, 22-23, 24-25, 26-27
  Metamorphic 7, 22-23, 24-25, 26
  Sedimentary 22-23, 24-25, 26-27, 29

S
Salt 5, 7, 8
Silver 5, 13, 14, 24

T
Talc 12-13, 30
Topaz 13

U
Uyuni Salt Flat, Bolivia 7

Z
Zircon 29